Evidence Collection Field Guide

Joseph J. Vince, Jr.

President, Crime Gun Solutions LLC
Frederick, Maryland

Professor of Criminal Justice
Mount St. Mary's University
Emmitsburg, Maryland

Bureau of Alcohol, Tobacco, and
Firearms (Retired)

William E. Sherlock

Chicago Police Department (Retired)
Chicago, Illinois

Illinois State Police (Retired)
Chicago, Illinois

JONES AND BARTLETT PUBLISHERS
Sudbury, Massachusetts
BOSTON   TORONTO   LONDON   SINGAPORE

World Headquarters

Jones and Bartlett
Publishers
40 Tall Pine Drive
Sudbury, MA 01771

Jones and Bartlett
Publishers Canada
6339 Ormindale Way
Mississauga, ON
L5V 1J2
Canada

Jones and Bartlett
Publishers International
Barb House, Barb Mews
London W6 7PA
United Kingdom

Copyright © 2006 by Jones and Bartlett Publishers, Inc.

Production Credits
Publisher, Public Safety: Kimberly Brophy
Acquisitions Editor: Stefanie Boucher
Associate Production Editor: Jenny McIsaac
Manufacturing and Inventory Coordinator: Amy Bacus
Text and Cover Design: Anne Spencer
Composition: Graphic World
Printing and Binding: Transcontinental
Cover Printing: Transcontinental

WRITING IN THE FIELD REFERENCE
For making permanent notes, use a felt-tip pen such as a Sharpie™ and allow the ink to dry thoroughly (may take as long as 30 minutes). For temporary notes, erase as soon as possible (within 10 minutes preferably) with alcohol.

Printed in Canada
09 08 07 06 05   10 9 8 7 6 5 4 3 2 1

Contents

Contents   iv

Contents   v

General Crime Scene Management

Crucial Precautions Upon Arrival at a Crime Scene

- Ensure that there are no imminent threats or hazards to the responders' health and safety or that of citizens in the immediate area.

- Identify and secure the boundaries of the primary crime scene by roping off or otherwise preventing unnecessary activity and traffic, including any area that is likely to yield evidence. We recommend using a double tapeline, using the standard yellow crime scene tape for the outside perimeter and red crime scene tape to denote areas accessible only to crime scene personnel. This also limits the biohazard contamination.

- Identify and protect secondary scenes, which are all avenues that could be used to travel to and from the primary scene.

- Do not allow unauthorized persons to enter the crime scene area. Establish only one point of entrance and egress.

- Prevent destruction or contamination of evidence.

- Photograph the crime scene to record its appearance when first responders arrived.

- Record the names and contact information of all citizens and officers who enter or leave the crime scene, as well as the time and the purpose of their visits.
- Record the weather and lighting conditions at the crime scene, and if they change, note the conditions and the time of change.
- Prohibit any unnecessary activity at the crime scene, including people walking through, touching surfaces, and moving or removing items.
- Do not use the telephone.
- Do not shut down or turn off computers.
- Do not use bathroom facilities.
- Do not allow food or drinks to be brought into the crime scene.
- Detain witnesses until investigators arrive, and separate them from each other to prevent their discussing observations with each other.
- Do not allow media persons into the crime scene; interviews and statements should be left to the Public Affairs Officer or a command officer.

Crucial Steps Required for the Collection of Evidence by First Responders

1. Determine who shall perform the scene processing.
2. Assess the scene to determine resources required.
3. Control contamination and prevent cross-contamination.
4. Document every item of evidence and action taken by personnel at a crime scene.
5. Prioritize the collection of evidence.
6. Collect evidence in the appropriate manner, as specified in later sections of this book.
7. Final survey, inventory, and debriefing should be performed before custody of the crime scene is released.

Word to the Wise

Public safety and liability mandate that each crime scene be cleared of any objects that may be hazardous. Make sure that each scene is clear of items and unsafe conditions that could cause harm to children, animals, or spectators who might wander onto the scene after your departure.

BEST PRACTICES

Professionalism 24/7 Makes for Successful Cases

Upon arrival at the scene, a plan must be developed that takes all variables into account. Examples of such variables could be outdoor lighting conditions, which might require portable lighting to be brought to the scene, or weather conditions, which might require a vehicle to be on hand for warming or cooling crime scene personnel.

Photographs, both over-alls and close-ups (with rulers or scales when required), should be taken of every piece of evidence, before packaging. Fill out your photo log/exhibits log before going to the next piece of evidence. Waiting until later may cause mistakes to happen.

Mark and seal your evidence uniformly and consistently each time. Train yourself to make sure the evidence is DICED—that is, marked with the following:

Date

Initials

Case number

Exhibit number

Description of the evidence and where it was recovered

Crime Scene Safety Measures

BEST PRACTICES

Basic Safety Procedures for the Suspected Presence of Hazardous Substances

- Properly ventilate the scene and/or use respirators when you suspect dangerous fumes are present. Chemicals or infectious agents can most easily enter the body through inhalation. The types of scenes where these items may be present include, but are not limited to, bombings, fires, chemical spills, derailments, natural disasters, and nuclear accidents. You should also take special care when arriving on the scene of a breaking and entering at places such as hardware stores (where there are pesticides, solvents, and other chemicals), pharmacies, chemical factories, petroleum distilleries, medical laboratories (legal and clandestine), mechanical garages, fireworks factories, and even retail establishments that may contain pesticides, solvents, or other chemicals.

- Wash hands immediately on departing the crime scene, especially before eating or smoking, to prevent ingestion of chemical or biological contaminants.

BEST PRACTICES—cont'd

- Wear gloves, shoe covers, and safety goggles when searching for and handling evidentiary material. Change gloves if they are torn, ripped, or punctured. Remove gloves when leaving the crime scene, and use new ones when returning.
- Place items in appropriate containers (see corresponding sections).
- Use full body suits when there is a possibility of exposure to spores of hazardous agents.
- Treat all body fluids, from living or deceased persons, as infectious. Allow body fluids to dry before packaging.
- Decontaminate all equipment after use. A solution of household bleach (diluted 1:10) or 70% isopropyl alcohol can be used to destroy contaminants. The bleach mixture must be prepared fresh because bleach loses potency over time. (Most crime laboratories make a new batch each day.)
- Affix the appropriate hazardous warning label to the outside of a package or container.

Physical evidence found at a crime scene may take many forms, including paint smears adhering to a criminal's vehicle that hastily flees the scene and strikes other vehicles or other objects that leave evidence (e.g., paint, plastic) on the vehicle. The importance of securing both the primary crime scene and the secondary scene cannot be overemphasized. The most commonly encountered types of evidence at armed assault and robbery scenes are fingerprints, fired bullets, discharged cartridge cases, body fluids (including blood), and binding materials.

When investigating document authenticity, consider whether any of the following circumstances are present:

- A signature that appears to be unnatural
- Paper that does not seem to be of the type customarily used for such documents
- Apparent differences in the types of ink used for such documents
- Use of more than one style of typewriting
- A questionable date of preparation

BEST PRACTICES

Responding Officer's Checklist for a Breaking and Entering Scene

- Identify and interview victim(s) and witness(es) and document the date, time, and location of interview.
- Identify who reported the crime, and interview that person or persons.
- Obtain personal identifiers from all interviewed.
  - Full name
  - Address
  - Telephone numbers (home and work)
  - Date of birth
  - Driver's license number or government picture ID
- Obtain full description of property lost. Acquire all identifiers, including make, model, serial number, and so on.
- Verify point of entry and exit.
- Verify method of attack.
  - Door pried
  - Window broken

continued

BEST PRACTICES–cont'd

- Roof penetrated
- Alarm system attacked, and so on
- Identify any tools used to gain entry.
- Systematically search the area for anything that may be of evidentiary value.
- Perform canvass of area.
  - Line of sight
  - Neighborhood
  - Roof, and so on
- Contact evidence technicians to collect evidence at scene, or self-perform.
- Direct evidence technicians to any additional evidence that you want recovered.
- Ensure that scene is photographed.

BEST PRACTICES

Checklist for First Responders to a Suspected Homicide Scene

- Immediately assess the victim's condition and determine whether aid can be administered. (Medical personnel must perform this examination and make this determination, because public safety officers are not qualified to make the rendering that the victim has expired.)
- Immediately call for medical assistance and notify the Coroner's office.
- Determine primary and any secondary scenes and follow the procedures outlined in the "General Crime Scene Management" section to fully protect the scene.
- Carefully search for and evaluate the presence of possible forensic evidence in order to notify forensic technicians and detectives.
- Photograph the scene and victim, as they existed at your arrival. (A log of photographs taken is essential for later investigation as well as for preparation of testimony.)
- If the scene is outside, take photographs of all persons (crowd) in the immediate area.

continued

★ BEST PRACTICES–cont'd

- Identify all possible persons to be interviewed and segregate them for interview by investigators.
- Make notes to ensure that you do not forget important information when interviewed by detectives and when preparing the report. If you detained or arrested an individual at the scene, it is important that all spontaneous utterances be documented and provided to detectives. *Every investigative action you take at the crime scene should be noted and documented in your report.*

■ Crime-Gun Process

The first step an officer should take when recovering a firearm is to ensure that it is rendered safe. That requires making sure that the weapon is safely unloaded and unable to chamber a round. Although it is desirable to have the firearm, its ammunition, and its magazine preserved for possible fingerprint recovery, the safety of fellow officers and citizens alike is paramount. Having made that statement, the firearm should be examined, unloaded, and cleared by the most qualified officer at the scene who is also familiar with the weapon.

BEST PRACTICES

Firearms Clearing Steps and Precautions

- Treat every firearm as if it were loaded.
- Always keep the muzzle pointed in a safe direction. (Use a firearm-clearing barrel where available.)
- Always keep your finger off the trigger and outside the trigger guard unless you intend to fire the weapon.
- Remove the magazine or source of ammunition.
- Open the breach and inspect the chamber to ensure that the firearm is completely unloaded. (*Note:* For some weapons, it is very difficult to discern whether the chamber is empty. Tubular magazines should have a BB or other such item passed completely through to establish proof of emptiness.)
- Use a wire cuff or trigger lock of some sort to ensure that the firearm cannot be reloaded or fired by mistake.
- *Never transmit or transport a loaded firearm to a laboratory or evidence storage facility!*

An automatic weapon (machine gun) is defined as any weapon that shoots, is designed to shoot, or can be readily restored to shoot more than one shot automatically, without manual reloading, by a single function of the trigger. Field testing a firearm to determine fully automatic fire capability requires an investigator to cycle the action of the firearm while holding the trigger depressed to the rear. The field-testing procedure is as follows:

1. *Make certain that the firearm is unloaded.* Remove the magazine, pull the bolt or slide to the rear, and visually inspect the receiver and chamber to ensure that the firearm is unloaded.
2. With the bolt or slide forward, pull the trigger to the rear and hold it in a depressed position to the rear while completing the next step.
3. Pull the bolt or slide completely to the rear and then release it (the bolt or slide should go forward to a closed position).
4. Release the trigger and then pull the trigger a second time.

a. If the firearm mechanism (pin) falls when the trigger is pulled the second time, then the field test indicates that the firearm is semiautomatic.

b. If the firearm mechanism (pin) does not fall when the trigger is pulled the second time, then the field test indicates that the firearm is fully automatic.

5. Place the safety or selector switch in the other available positions and repeat the function test and field test. The firing mechanism should not function with the safety or selector switch in the "Safe" position. If it does, the firearm may be capable of fully automatic fire.

6. Qualified firearm experts (designated by the respective investigative agency) who have the credentials to testify to the firearm's capability in court should additionally test any firearm suspected of being capable of fully automatic fire.

A short-barreled rifle or shotgun is defined as (1) a shotgun having a barrel or barrels of less than 18 inches in length; (2) a weapon made from a shotgun if such weapon, as modified, has an overall length of less than 26 inches or a barrel or barrels of less than 18 inches in length; (3) a rifle having a barrel or barrels of less than 16 inches in length; or (4) a weapon made from a rifle if such weapon, as modified, has an overall length of less than 26 inches or a barrel or barrels of less than 16 inches in length.

To determine the barrel length of a rifle or shotgun:

1. Examine the firearm and *ensure that it is unloaded*.

2. Close the breech, breechblock, cylinder, or bolt.

3. Cock the firearm to withdraw the firing pin.

4. Insert a straight rod down the muzzle end of the barrel until contact is made with the face of the bolt, breech, or breechblock.

5. Mark the rod at the muzzle end (the furthest extent of the barrel) to denote the true barrel length (see Note).

6. Withdraw the rod and measure the distance marked on the rod.

To determine the overall length of a rifle or shotgun:

1. Examine the firearm and *ensure that it is unloaded*.

2. Close the action of the firearm if it is a break-open type of action.

3. Lay the firearm on its side on a table with the butt of the stock in line with one edge of the table. Keeping the butt against that one edge, bring the barrel of the weapon in line with (parallel to) and next to the right-angle edge of the table.

4. The overall length of the firearm can then be determined by measuring the distance of the right-angle table edge from the beginning, the butt of the firearm, to the point where the end of the barrel or barrels is located (see Note).

Notes

Note: Removable barrel extensions, poly chokes, flash suppressors, and so forth, are not considered to be part of the measured barrel length; however, permanently affixed attachments are considered part of the barrel and are therefore included in the measured barrel length.

BEST PRACTICES

Checklist for Tracing Firearms

In order for ATF to conduct a firearm trace and obtain information from federally licensed manufacturers, importers, and dealers, it is essential that ATF receive a complete and accurate description of the firearm. Law enforcement officers and investigators must ensure that the description of the requested firearm to be traced is complete and accurate.

- Firearms cannot be traced without a serial number, although at times they may be traced with a partial serial number. Ensure that any letters adjacent to the serial number are also recorded on the trace form. In the case of a partial serial number, contact the National Tracing Center (NTC) for guidelines.
- Most firearms cannot be traced without a model. Therefore, always include the model on the trace form.
- Foreign-made firearms cannot be traced if the importer's name is not reported. Licensed importers are required to mark the firearm with their name, city, and state, as well as the country of origin. In many

BEST PRACTICES—cont'd

cases, this information appears on the barrel in an abbreviated form.

- Surplus military firearms are generally untraceable unless the firearm is marked with an importer's name. Many foreign and domestic surplus military firearms have been imported in recent years and can be traced if the importer's name is reported on the trace form.

- ATF Form 3312.1, Crime Gun Information Request/ Referral Form, should be used to request a firearms trace. It is important that the information on the form be as complete as possible.

SECTION

2

Collection and
Preservation
of Evidence

■ Arson Debris

Collect carpet, wood, and other absorbent materials located close to the suspected origin of the fire as comparison standards. Use the packaging procedure described for glass fragments.

■ Body Fluids and DNA

- Body fluid stains are valuable evidence. They can be used to associate a suspect with a crime or to eliminate a suspect from consideration. Because of the dangers in handling whole blood, blood collection cards are being widely employed. If a blood standard is needed from the morgue, a blood collection card is used: Several drops of blood are put on the card and allowed to dry. When the blood is dry, it is packaged and DICED: dated, initialed, case numbered, exhibit numbered, and a description of the evidence recorded.

- Unlike whole blood, there is no need for refrigeration of dried blood. The possibility of having broken glass tubes, spillage, or spoilage due to improper refrigeration is eliminated with the use of blood collection cards.

- Precautions must be taken not to contaminate the samples by coughing, sneezing, or even talking while collecting this type of evidence. With the continuing growth in the field of DNA, contamination at the crime scene is an increasing problem.

- The sensitivity of the automated systems in use today requires that personnel at crime scenes wear personal protective equipment (PPE). Unprotected clothing can shed head hair and dead skin cells that contaminate the scene. Not taking proper protections can result in obtaining false leads and therefore wasting costly investigative time.

- After the first responders have turned the scene over to laboratory personnel and detectives/investigators, all law enforcement personnel should be required to wear PPE when entering the crime scene.

BEST PRACTICES

Detectives must be dutiful in changing gloves to prevent cross-contamination any time that they touch an object at the crime scene. The same policy should apply to any shoe coverings that are worn.

Avoiding Contamination

- During the evidence collection process, contamination can also occur when collection tools are not cleaned after picking up an item. In this way, DNA evidence may be transferred from item to item.

- If possible, disposable equipment, such as tweezers and forceps, should be used. This prevents contaminating another scene with DNA from a prior scene and creating a false serial pattern.

- Equipment that is nondisposable (e.g., cameras, tripod) should be carefully cleaned before being returned to a carrying case or vehicle.

- Clean, sterile paper goods (envelopes, bags, and boxes) should be used for packaging DNA evidence. Items must be thoroughly dried before packaging; this prevents mold, moisture, and sunlight from degrading biological evidence.

- Items that cannot be dried at the scene can be placed in a bag or envelope and properly sealed and identified. This package can now be placed in an *open* plastic bag for transportation. This process contains any leakage, which could spread contamination and biohazards.

- Evidence should be kept in a cool, dry atmosphere or refrigerated until it can be taken to the laboratory.
- The plastic bags used for transporting the evidence can be properly disposed of when the evidence has been secured in the drying station. These bags are not part of the chain of evidence; they are used only to transport a potentially biohazardous material until it is placed in a safe environment.

The following sections describe procedures for collecting and preserving blood, saliva, semen, and urine, as well as comparison standards where applicable and possible.

Blood

Materials stained with blood that can be sent to the laboratory should be handled in the following way:

- Set the bloodstained material on a piece of clean paper in a draft-free area and allow it to air-dry.
- Place the dried material in a paper bag; mark the package with a date, initial, case number, exhibit number, and description of the evidence (DICED); and seal the bag. Any debris that falls from the material onto the paper during the drying process should be placed in a separate

paper envelope; marked with the date, initial, case number, exhibit number, and a description of the evidence; and sealed.

- If you must fold the material, protect the stained area with a piece of paper. Folding an unstained portion of the garment onto a stain may cause a stain transfer.
- Wrap each bloodstained item separately.

Word to the Wise

Do not package items when they are still moist. Allow them to dry thoroughly.

Word to the Wise

Photograph all areas before any evidence collection takes place.

Items stained with blood that cannot be sent to the laboratory should be handled as follows:

- Scrape the dried blood onto a clean piece of paper, using a clean knife or razor blade. (Wipe the packaging matter from new razor blades before use.)
- Similarly scrape the nonstained area immediately surrounding the stained area onto another piece of paper.
- Fold each paper, firmly securing the scraping, and place each paper in an envelope.
- Write the date, initials, case number, exhibit number, and description on the envelope, and seal it.
- Package and preserve the razor blade.

When quantities of wet blood are available, use the following procedures:

- Soak up the moist blood with a gauze pad.
- Air-dry the gauze pad.
- Place the gauze pad in a paper envelope, seal it, and label the envelope with the date, your initials, case number, exhibit number, and description of the evidence.
- *Do not cover vent holes in the box or completely cover the envelope with tape.*

Collection of Comparison Standards for Blood
Collection of comparison standards for blood is best accomplished with a Buccal Swab Collection Kit. Check with your forensic laboratory to see whether it has a preference for the brand of Buccal Swab Collection Kit to use.

Perspiration
- Use a Buccal Swab Collection Kit.

Saliva
- Use a Buccal Swab Collection Kit

Collection of Comparison Standards for Saliva
See the previous section for information on collecting saliva.

Semen
- Use a Buccal Swab Collection Kit.

Urine
- Use a Buccal Swab Collection Kit.

■ Cigarette Butts and Tobacco
- Pick up the cigarette butt on a piece of paper or with tweezers and place in a small paper bag. *Do not handle the cigarette butt directly with your hands.*

BEST PRACTICES

Cigarette or cigar butts should be submitted for DNA comparison.

- Mark a label with the date, initials, case number, exhibit number, and description, as well as where the object was found.
- Place the label on the plastic container and seal.
- Empty tobacco material from pipes or clothes' pockets into a plastic-type pillbox. Mark and seal as described for cigarette butts.

■ Documents

Charred Documents
- Carefully place charred documents between layers of paper.
- Enclose the documents in a crush-proof container, seal the container, and label it with your initials, the date, and a case and exhibit number.
- Indicate "FRAGILE" on the mailing container.

Computer Files

- Note the type and brand of computers and operating systems.
- If the computers are networked, determine the type of network software used, the location of the network servers, and the number of computers on the network.
- If possible, ascertain whether encryption and/or password protection is used.

Unless an examiner advises otherwise, most investigations require only the central processing unit and external storage media (e.g., hard drives, CDs, floppy disks, tapes). Equipment collected for transport or shipping should be placed in sturdy cardboard containers. If possible, try to use the original or other computer packaging with fitted padding. Pack the equipment with bubble wrap or other similar padding to avoid jarring. Do not use materials that can break into small pieces (e.g., Styrofoam) and lodge inside the processor and damage internal parts. Affix an evidence tag with the date, your initials, a case number, an exhibit number, and a description of the contents. Place a transmittal document inside the box as well as affixing one to the outside. Seal the box with heavy packing tape. Always ship processing units upright and mark the container "THIS END UP." Label the outer container as follows: "FRAGILE—SENSITIVE ELECTRONIC EQUIPMENT. KEEP AWAY FROM MAGNETS OR MAGNETIC FIELDS."

CDs, floppy disks, tapes, or hard drives should be packaged separately and tightly to avoid movement during transport or shipping. With a marker, date and initial the container, and add a case and exhibit number. As with the processor, label the outer container as follows: "FRAGILE—SENSITIVE ELECTRONIC EQUIPMENT. KEEP AWAY FROM MAGNETS OR MAGNETIC FIELDS."

Crumpled Documents
- Follow the procedures outlined in the section "Charred Documents."

Handwriting Samples
To obtain a handwriting sample, after having informed the suspect of his or her constitutional rights, request the subject to write the text of the questioned material several times.

Do not allow the suspect to view the questioned writing, and give no instruction as to spelling, punctuation, and so forth.

In forgery cases, obtain samples of the signature of the person whose signature was forged. If the text of the questioned material is extensive, have the subject write a substantial part of it each time.

- Make sure that the examples are written on something other than the questioned document and are witnessed at the time of writing by you or by another person.
- If possible, try to make the example comparable to the questioned document in the type of writing instrument used, the kind of writing surface used, the amount of space available for writing, and the type of writing (i.e., cursive, printed, and so on).
- Also, in the case of a legal document in which the age of the document is in question, try to obtain other documents executed near the date of the questioned document.

- Clearly label the samples as standards and submit them along with the original document in a protective covering, such as a large envelope or plastic covering, that is sealed and labeled with your initials, the date, and a case and exhibit number.

Ink Examples
- Collect samples of unquestioned documents that contain the same type of ink that was used in the questioned document. If you are unsure whether the ink on the standard is the same as that on the questioned document, submit samples of all available inks that might have been used. If samples of fluid ink are available, place the bottles containing the ink in a suitable container that is labeled with the date, initials, case number, exhibit number, and description; indicate "FRAGILE" on the container; and submit it to the laboratory.

Intact Documents
- Whenever possible, submit the original document rather than a photograph, photostat, or other type of copy.
- Handle the document carefully, using tweezers or gloves as necessary; preserve latent fingerprints.

- When identification is necessary, mark in a noncritical area of the document. Use a medium different from that used on the document (e.g., use pencil when the document is in ink). *Do not use staples or pins on the documents. Do not fold the documents.*
- Place the document in a protective covering, such as an envelope or plastic covering, and then seal and label the envelope with the date, initials, case number, exhibit number, and description.
- Collect a comparison standard.

Typewriter and Printer Examples
- Collect specimens of typewriting or printing from the typewriter or printer believed to have been used and submit them to the document analyst. When obtaining these examples, use the typewriter or printer in the condition in which it is found (e.g., with the same ribbon or toner cartridge). Type the text of the questioned material three times (for the typewriter, type once through carbon paper with the ribbon adjustment set on "stencil"). On each specimen, record the brand name, model number, and serial number of the typewriter or printer that was

used. Place specimen papers taken from the typewriter or printer in a plastic or reinforced envelope for storage and transmissions. Do not fold the specimen papers; label the envelope the date, initials, case number, exhibit number, and description.

■ Explosive Debris

- To collect debris left after an explosion, first locate the seat of the blast. Remove the first half-inch of soil or debris from the crater and proceed as follows: place the sample in a sealed glass jar or metal can and label it with the date, initials, case number, exhibit number, and description.

- Place objects thought to have been in close proximity to the blast in a sealed glass jar or metal can and label. Soft materials, such as cloth, rubber, and other readily penetrated materials, are good collectors of explosive residues. Collect and preserve them. Sharp objects should be packaged in metal containers.

- Place metal fragments from the explosive device in a sealed glass jar or metal can and label it with the date, initials, case number, exhibit number, and description.

- When packaging wire or other objects with tool marks, protect the items by wrapping them in tissue paper secured with tape. *Do not put tape directly on the tool mark area; valuable trace evidence could be lost when the tape is removed.* If for some reason items such as wire or rope must be cut in order to be collected, mark the end or area that was cut for recovery. Sharp objects should be packaged in metal containers.

- If you find a suspected dynamite wrapper, place it in a sealable glass container and label with the date, initials, case number, exhibit number, and description. If a glass container is not available, use a metal can. *Do not pack the dynamite wrapper in the evidence box near materials taken from a suspect.*

Word to the Wise

Do not allow sharp-edged objects to pierce the sides of a bag. If sharp-edged objects are present, place them in a metal container.

■ Explosive Substances and Devices

> **Word to the Wise**
>
> Never attempt to deactivate any explosive device yourself unless you have been trained to perform that specific function!

When dealing with explosive substances and devices, be sure to collect evidence only from deactivated devices. Process smooth surfaces of the device for latent fingerprints. Place tape, paper wrappings, or any materials that may bear latent fingerprints in a paper envelope, and label it with the date, initials, case number, exhibit number, and description. *Do not handle any of these items directly with your fingers.*

- Separate all components, such as blasting caps, batteries, and wires.
- Place each component in a separate plastic bag, making sure that all items bearing tool marks are wrapped in tissue paper and secured with tape.

- Seal each bag and label the date, initials, case number, exhibit number, and description.
- Place a small quantity (1/8 teaspoon) of the suspected explosive in a small sealable glass or metal container. *Do not send explosives through the mail.*

■ Explosives and Incendiaries

Care should be taken to ensure that cross-contamination does not occur. Place evidence only in containers that have not held other items or that have been properly sterilized before reuse. Caution should also be taken not to use plastic containers because they are made of a petroleum base and may cause cross-contamination. Packaging device components in airtight containers can reactivate explosive materials (e.g., MacGyver bombs) and should be avoided.

Glass Fragments
- If it appears that a Molotov cocktail type of device was used, collect glass fragments and dust them for fingerprints.
- Place glass fragments in a sealable metal container (sterile friction-top cans, paint cans, or screw-cap glass jars, such as Mason jars, are excellent for this purpose), and

mark the container with a date, your initials, a case number, an exhibit number, and a description of the evidence. Such containers can be purchased from can manufacturers or forensic supply houses.

- Seal the container tightly so that any vapors present cannot escape.

  Also see the information in the "Glass" section.

■ Flammable Liquids

- Place 1 ounce (or whatever quantity is available) of suspected flammable liquid in a small glass bottle with a tight-fitting cap.
- Seal the container and mark with the date, initials, case number, exhibit number, and description.

■ Foreign Objects

- Place burnt matches, wire, or other objects apparently foreign to the scene in separate plastic or paper envelopes.
- Seal the package and mark it with the date, initials, case number, exhibit number, and description.
- Collect a comparison standard.

■ Fabrics

Any and all fabric found near or at the crime scene, or found to be missing from the crime scene, may be either supportive or necessary evidence in establishing a relationship between the crime scene and the suspect. Collect and preserve such items carefully to avoid contamination.

Large Articles

Before collecting and packaging large articles, such as mattresses and upholstered chairs, record the exact position of the evidence. For example, indicate in your notes or diagram which end of a bloodstained mattress was next to the headboard. Be especially careful during the collection and preservation process, so as not to loosen any trace materials.

- Do not tear, stretch, or handle fabrics roughly.
- Allow wet surfaces to dry before packing evidence away.
- Carefully fold large pieces of fabric, protecting any torn edges, and place in a clean bag, which should then be sealed and marked.
- *Plastic bags and bottles are not satisfactory for packaging any material that may contain petroleum.*

- Package large articles, such as mattresses and upholstered chairs, intact in large crates or boxes.
- Affix an evidence tag to all articles and mark the date, initials, case number, exhibit number, and description.
- Collect a comparison standard if possible.

Small Articles

Look carefully for small articles of fabric throughout the general crime scene, at entrances and exits, and on any victims. *Do not overlook fibers that may be on the victim's mouth, feet, or hands.* Search for imprints of fabric weave in painted surfaces, in putty, or on other objects. Handle these carefully; do not contaminate with fingerprints or other impressions.

- Collect small articles carefully with tweezers, ensuring that torn edges are protected.
- Allow small articles containing wet surfaces to dry before packing. These surfaces should be protected with nonabrasive material during shipment.
- Place evidence in small containers, but not so small as to require folding of the evidence. Folding causes distortion of threads.

- Fabric impressions on objects such as paint, metal surfaces, and putty require careful handling. See the sections "Hair and Fibers on Hard Surfaces," "Tool Marks," and "Chips and Smears" for collection and preservation procedures.
- Package each fabric impression separately in a glass or plastic vial, a small box, or other appropriate container.
- Seal each container, and label with the date, initials, case number, exhibit number, and description.
- Collect a comparison standard.

■ Fingerprints

Fingerprints on Absorbent Materials

- Place the paper or other absorbent material in a plastic bag or cellophane protector. *Do not handle the material with your fingers. Do not attempt to develop latent fingerprints on absorbent surfaces yourself.*
- Label the bag or protector with the date, initials, case number, exhibit number, and description.
- Collect a comparison standard.

Fingerprints on Hard Surfaces
- Dust plastic cards, metal plates, glass bottles, or other hard-surfaced objects for latent fingerprints.
- Remove developed prints with lifting tape and place the tape on a 3-inch by 5-inch card that contrasts in color with the dusting powder used.
- Mark the card the date, initials, case number, exhibit number, and description, and seal the envelope.
- Collect a comparison standard.

Fingerprints on Soft Surfaces
- Carefully remove putty, caulking compound, or other soft material bearing visible fingerprint impressions. Leave as much excess material surrounding the fingerprint as possible.
- Glue the mass of material to a stiff section of cardboard that is marked with your initials, the date, and an exhibit and case number.
- Tape a protective cover over the specimen. A paper cup or baby food jar can be used for this purpose. *Do not touch or otherwise distort the fingerprint.*
- Collect a comparison standard.

■ Firearms and Ammunition

Firearms leave unique markings on fired bullets and discharged cartridge cases as well as detectable gunshot residue on the shooter's hands.

Word to the Wise

When handling a firearm, due care should be taken to ensure that the firearm is not loaded. Every firearm should be considered loaded, and a trigger lock or other device should be used when examining firearms.

BEST PRACTICES

Dos and Don'ts for Handling Firearms

Do

- Store firearms in paper envelopes, paper bags, or cardboard boxes. Plastic collects moisture, which causes firearms to rust. Commercial heavy-duty paper envelopes, paper bags, and cardboard boxes of all sizes

continued

BEST PRACTICES—cont'd

that are specifically designed for weapons storage are available through law enforcement supply vendors.

- If the recovering officer is not sure that the firearm has been rendered safe, the firearms section of the laboratory should be contacted so that arrangements can be made to hand carry the weapon. If the laboratory is located at a distance from the scene, the department's firearms instructor or range master should be contacted for assistance.

Do Not

- *Never send a loaded firearm through the mail.* Federal law and postal regulations forbid transmitting firearms in this manner.
- Do not mark the firearm in any permanent way. In past years, officers have scratched their initials on seized firearms for later court identification. The weapon in question could be a valuable collector's piece, and scratching initials into the finish of the firearm could reduce its value to the owner. Departments have been held liable for damages to firearms that were not properly stored and received a minimum of care.

Bullets

- Remove fired bullets from any object, leaving a layer of extraneous material surrounding the bullet. The firearm examiner should remove any plaster residue or other foreign material.
- Place the fired bullet in a paper envelope.
- Label the container with the date, initials, case number, exhibit number, and description.
- *Do not scratch the outer surface of the bullet while removing it from an object.*
- *Do not scratch identification marks on the sides or base of the bullet.*

Discharged Cartridge Cases

- Do not mark the discharged case or cases for identification.
- Place the cartridge case or cases in a paper envelope.

  Seal the package and list the date, your initials, and corresponding case and exhibit numbers.

Gunshot Residue

Gunshot residue evidence collection kits contain three stubs that have an adhesive tape surface. Two of the stubs are used to press on the back of the suspect's left and right hand in the area of the index finger and the thumb. The third stub is used to take a sample of the atmosphere in the area that the test is being performed. Clothing may also be examined for these components with the use of these kits. If the presence of gunshot residue is in question, care must be taken when submitting clothing. Collect and transmit each garment in separate paper bags. It is also suggested that a small paper bag be placed over the cuff end of the sleeves to about the elbow area, which helps prevent any transfer of the residue.

Handguns
Pistols
- Unload the pistol, remove the magazine from the weapon, and check the chamber.
- Dust the pistol, the exterior of the magazine, and any cartridges left in the magazine, making note of the number of cartridges left in the magazine. *If a discharged cartridge case or cartridge is found in the pistol's chamber, place it in a separate envelope and make a note for the firearms examiner.*
- Ensure that you close and lock the action of the weapon.

- Attach an identification tag to the weapon that describes the pistol and lists its make, model, caliber, importer (if any), finish, and serial number. If the serial number is obliterated, note that on the tag. List on the tag the date, initials, case number, exhibit number, and description.
- Always put cartridges in an envelope and seal securely when they are being packed with a weapon.
- Place the weapon and the sealed envelope containing the cartridges in another envelope and seal. Mark the envelope the date, initials, case number, exhibit number, and description.

Revolvers
- Unload the revolver, taking note of the position of discharged cartridge casings and/or live cartridges in the cylinder with respect to the barrel.
- Close and lock the cylinder on the revolver.
- Process the weapon, cartridges, and discharged cartridge cases for fingerprints.
- Attach an identification tag to the revolver that lists its make, model, caliber, importer (if any), finish, and serial number. If the serial number is obliterated, note that on

the tag. List on the tag the date, initials, case number, exhibit number, and description.

- Always put cartridges in an envelope and seal it securely when the cartridges are being packed with a weapon.

- Place the weapon and the sealed envelope containing the cartridges and discharged cartridge cases in an envelope and seal (see Word to the Wise). List on the envelope the date, your initials, the corresponding case number and exhibit number, and a description of the contents.

Shoulder Weapons

- Unload the weapon, remove the magazine from the weapon, and check the chamber. Weapons with tubular magazines should be carefully examined to ensure that no cartridges or shot shells are stuck in the tube.

- Dust the exterior of the weapon, the magazine, and the cartridges left in the magazine, and make a note of the number of cartridges left in the magazine. *If a discharged cartridge case or cartridge is found in the chamber, place it in a separate envelope and make a note for the firearms examiner.*

- Close and lock the action.

- Affix an identification tag to the weapon, describing the weapon and listing its make, model, importer (if any), caliber, finish, and serial number. Note whether the weapon is fitted with a scope and/or scope mounts, shoulder sling, bayonet mount, or fixed bayonet. List on the tag the date, your initials, the corresponding case number, and an exhibit number.
- Always put cartridges in an envelope and seal securely when they are being packed with a weapon.
- Place the unloaded weapon and sealed envelope containing the cartridges in a wooden or sturdy cardboard box. Seal the box; list the date, initials, case number, exhibit number, and description.

Weapons with Obliterated Serial Numbers
- Make an identifying mark on the weapon if it has an obliterated serial number. (The weapon is already damaged and most likely would not be returned to anyone.)
- Affix an evidence tag to the weapon containing the date, initials, case number, exhibit number, and description.
- Place the weapon in a paper bag.
- *Do not attempt to recover the number yourself with acid etching solutions.*

■ Food and Drug Specimens

Liquids

- Try to collect a minimum of 1 pint of the specimen. Use a leak-proof container for this purpose.
- Seal the container with adhesive tape, and label it with the date, initials, case number, exhibit number, and description.
- If the container used is glass or has a glass stopper, mark it "FRAGILE."
- Collect a comparison standard.

Word to the Wise

It is extremely important to collect the victim's vomit for examination and analysis. Vomit should not be overlooked or washed away.

Plant Material

- Thoroughly dry the sample by spreading it on brown paper for at least 24 hours.

- After the sample has been dried, place it in a pillbox, vial, or other container, and secure the container with adhesive tape. *Do not mix samples; package each separately to avoid mixing during mailing.*
- Label the outside of the container with your initials, the date, and an exhibit and case number.
- Collect a comparison standard.

Powders or Solids
- Place the powder or solid in a container, such as a pillbox or plastic vial.
- Seal the container and label it with the date, initials, case number, exhibit number, and description.
- Refrigerate samples as needed. *Do not add preservatives to solid food samples.*
- Collect a comparison standard.

Tablets and Capsules
- Place the tablets or capsules in a container, such as a pillbox or plastic vial.
- Seal the container and label it with your initials, the date, and an exhibit and case number.

- Clandestine factory equipment leaves tool mark impressions; have pills checked for tool marks from a pill press. In addition, check tool marks with other known suspect pills for a possible drug trafficking pattern, including locations, associates, and financial partnerships.
- Collect a comparison standard.

■ Glass

Large Fragments
- Dust fragments for latent fingerprints and submit prints.
- Protect thin protruding edges of fragments against damage by embedding them in modeling clay, putty, or any similar substance. *Avoid chipping the fragments.*
- Use tweezers or other similar type of tool to collect glass. Exercise care to protect the edges and avoid scratching the surface.
- Place adhesive tape on each piece for identification. Place the date, initials, case number, exhibit number, and description..
- Wrap each piece separately in paper and place them in a sturdy box with a tight-fitting lid.

- Seal and label each package with the date, initials, case number, exhibit number, and description.
- Package questioned pieces of glass separately from known pieces of glass.
- If you are submitting glass for the purpose of determining the direction of a bullet's impact or for any other fracture analysis, mark surfaces with tape indicating whether the glass was found outside or inside the building. Likewise, mark glass taken from a window frame and show which side was facing outside.
- Collect a comparison standard.

Small Fragments
- Examine articles of clothing and shoes for the presence of glass fragments.
- Use tweezers or other similar tool to collect glass. Use care to protect the edges and avoid scratching the surface.
- Wrap each article of clothing containing fragments separately in clean paper or plastic bags.
- Package any questioned pieces of glass separately from known pieces.

- Seal each bag and label each with the date, initials, case number, exhibit number, and description.
- Place shoes and other solid objects in separate containers, such as shoeboxes. Tape each object to the bottom of containers to prevent rattling. *Do not pack articles containing microscopic fragments in cotton or other soft protective materials.*
- Seal each package completely, making sure that there are no holes through which glass fragments might be lost.
- Place loose glass fragments in pillboxes or in plastic or glass vials and seal them tightly. Place paper in the container to prevent rattling and chipping during transit. *Do not use envelopes as containers.*
- Label everything with the date, initials, case number, exhibit number, and description.
- Collect a comparison standard.

■ Hair and Fibers

Hair and Fibers on Fabric Surfaces
- If hairs or fibers are observable, remove them from the surface with clean tweezers.

- The best method for capturing relevant case evidence from a fabric is to use an adhesive tape to lift recent debris that has been dropped onto the fabric. After lifting the material with tape, fold the tape over onto itself so that it does not adhere to the envelope. This is many times preferable to vacuuming because that method pulls in debris that may have been embedded in the fabric for a long period of time.

- If you use an evidence sweeper (vacuum), remove the material that accumulates in the filter and the filter paper and place it in an evidence bag. Seal the evidence bag and label it with the date, initials, case number, exhibit number, and description.

- *Do not use envelopes for packing filter sweepings or for other very small materials, and do not crush hairs with tweezers.*

- Collect a comparison standard.

Hair and Fibers on Hard Surfaces

- Remove suspicious hairs or fibers from surfaces with clean tweezers.

- Place these items in a pillbox or a folded piece of clean paper. If you use paper, avoid kinking the hair or fiber when folding the paper. It is recommended that you seal it in an envelope.

- Seal and label the container with the date, initials, case number, exhibit number, and description.
- *Do not crush hairs with tweezers when handling them.*
- Collect a comparison standard.

Hair and Fibers on Individuals

- When it is desirable to obtain hair samples from an individual, pull or comb out about 50 hairs with a clean pair of tweezers or a clean comb. A less desirable method for obtaining hair samples from an individual is to cut them off as close to the skin surface as possible. However, cutting or clipping gives only part of a sample, and misleading information can result if the hair has been recently dyed.
- Obtain samples of hair, whether pulled, combed, or cut, from various locations (arm, underarm, chest, pubic, and leg areas), and package them separately according to their location. Remove hairs surrounding multiple wounds on a body and place them in a container that is marked with the date, initials, case number, exhibit number, and description.

- If a pillbox is used as a container, ensure that the entire length of the hair fits in the box and that it is not crushed between the top and bottom when a lid is placed on it. If paper is used, avoid kinking the hair or fiber when folding the paper. Mark the box or paper with the date, initials, case number, exhibit number, and description.

- Also submit articles of the individual's clothing. Package each article separately in evidence bags, and then seal and label the bags with the date, initials, case number, exhibit number, and description.

■ Impressions

Footwear Impressions

Photograph each print that you have located. Follow these steps:

- Place a ruler or other means of scalar identification next to the print.

- Take the photograph by holding the camera directly over the impression (the use of a tripod is preferable whenever possible) while illuminating the impression by holding a detached flash or strobe light low and to the side of the impression.

2

- Prepare a dental stone cast of the impression of each footprint. *It is a good idea to practice first on your own footwear impressions because casts destroy impressions when they are removed.*
- When the cast is dry, scratch the date, initials, case number, exhibit number, and description in the cast.
- Also, scratch or mark a directional arrow pointing north on the back of the cast.
- Wrap each cast separately using corrugated paper and place in a well-padded box or container. *Do not remove dirt adhering to the cast.* Use torn newspaper or excelsior for padding.
- Seal and identify the contents of the box with a label.
- Collect a comparison standard.

Tire Impressions
- Follow the same procedures as described in the "Footwear Impressions" section to collect a tire impression.
- Collect a comparison standard.

Tool Marks

- Whenever possible, preserve tool marks as you find them and submit the intact object bearing the tool marks to the laboratory.

- If it is not possible to submit the intact object bearing the tool marks to the laboratory, remove that portion or section of the object bearing the tool mark and submit it to the laboratory (e.g., fender of car, door jambs). Affix an evidence tag labeled with the date, initials, case number, exhibit number, and description.

- Before a portion or section is removed from a large item, photograph the entire item. Submit this photograph with the intact object being forwarded to the laboratory for examination. Mikrosil casting material (or a comparable product) may be used for tool mark evidence that cannot be sent to the laboratory.

- Collect a comparison standard.

■ Jewelry

- Handle jewelry with tweezers or cloth gloves.
- Dust for fingerprints and place the item in a suitable crushproof container.

- If the composition of precious metals, such as gold, silver, or platinum, must be determined in order to prove common origin, send appropriate metal samples for comparison purposes.
- Label each sample container by writing on it in ink the date, initials, case number, exhibit number, and description.
- Collect a comparison standard.

■ Liquids and Viscous Substances

Liquids

- Handle containers with cloth gloves.
- Dust containers holding liquids for latent fingerprints. Process and submit any prints that appear.
- Pour the liquid into glass bottles no larger than vial size (2 ounces) with tight-fitting, narrow-neck screw caps. If the cap is metal, make sure that it is lined with inert plastic to prevent contamination of the liquid.
- If the liquid has been spilled, spoon or scrape as much of the liquid as possible into a tightly capped glass bottle. Even if only a few drops can be obtained, a meaningful analysis can be performed.

- If the liquid has been spilled onto a porous surface, such as carpeting or soil, place the wet portion of the material into a tightly capped glass or metal container.
- Pack the bottles carefully and surround them with foam rubber, polyurethane chips, or similar packing material.
- Make sure that each sample container is identified in ink with the date, initials, case number, exhibit number, and description.
- Label the packing boxes "FRAGILE."
- If a liquid is suspected of being hazardous (e.g., flammable), consult postal authorities for the correct mailing procedures.
- Collect a comparison standard.

Viscous Substances
- Handle containers with cloth gloves.
- Dust containers for latent prints and submit any prints that appear.
- If the substance appears to be grease, handle it in the following manner:

- For small amounts, wipe up the substance with a cotton swab, place the swab in a labeled plastic bag, and seal. Small smears may be sufficient for analysis.
- For large amounts, transfer the substance to a sealable plastic container using cotton swabs or a plastic spoon.
- Seal and label the container.
- If the substance appears to be oil, ink, or glue, do the following:
  - If possible, transfer the substance by pouring it into a glass or plastic wide-neck bottle with a tight-fitting screw cap.
  - If the material is in a tight-fitting container that is not larger than 1 pint, submit it in the original container.
  - If the substance is too viscous for pouring, transfer it to a container using a clean plastic spoon or other instrument.
- Package all exhibits as described earlier for liquids, making sure that each exhibit container is identified in ink with the date, initials, case number, exhibit number, and description.
- Collect a comparison standard.

■ Locks and Keys

Locks should be examined for the following:

- Examine locks for signs of picking. Also, when submitting locks to the laboratory, send a key for each lock. This aids in the disassembly of the lock so that microscopic examination can be conducted.

- Examine locks for delay devices that may have been used. For example, before making a roof entry into a store, a burglar may put delay devices on all the entrance doors. This prevents entry with a key and alerts the burglar so that he or she can make an escape from another door. Superglue and toothpicks are some of the things that can be put into the keyway of a lock as a delay device. Determining that a delay device was used could help detectives in establishing a crime pattern.

- Determine whether the lock is part of a master system.

- Determine the operating condition of the lock.

■ Magnetic Tape Recordings

When making recordings of anonymous phone calls, follow this procedure:

- Eliminate all background noise when making recordings.
- Use polyester or Mylar-backed tape that is at least 1 millimeter thick.
- Use standard cassette recorders or reel-to-reel recorders that are either electrically or battery operated. Make sure that batteries are fully charged.
- If you use cassette tapes, use C-15, C-30, C-60, or C-90 cassettes. *Do not use mini-cassette recorders*.
- Use only new or bulk-erased tapes.
- Use an inductive pickup, such as a donut or suction cup pickup.
  - If a donut pickup is used, place it over the earpiece on the telephone and plug it into the microphone jack on the recorder.
  - If a suction cup is used, place it on the backside of the telephone handset.
- Identify the tape recording container by writing in ink on the container the complaint number, date, time,

location, telephone number, and name of investigating officer.

- Before packing the tape, place it in its original box and wrap the box with aluminum foil to guard against magnetic fields, which may alter or destroy the recorded material.
- Collect a comparison standard.

Collection of Comparison Standard for Magnetic Tape Recordings

Make recordings of the voices of suspects so that a comparison can be made to the recording of the original anonymous voice. Use the following procedures:

- Duplicate the recording conditions of the anonymous call as closely as possible.
- If the original recording was made over the telephone, make the comparison recording over the telephone.
- Have the suspect repeat the exact text as the anonymous call.
- Ask the suspect to speak in his or her normal voice, but notice whether the suspect appears to be disguising his or her voice.

2

- Package and identify comparison recordings as described earlier in "Magnetic Tape Recordings."

■ Metals

Chips and Smears
Small Portable Objects and Clothing Containing Paint Chips or Smears

- Mark your initials in an inconspicuous place.
- Place each item in a separate envelope or paper bag, making sure that the area containing the paint is protected from any abrasion or destruction; cover these areas with plastic or brown paper.
- Pick up paint chips either by using tweezers or by scooping them up with a piece of paper. Place paint chips in a bag.
- Seal and label each bag with the date, initials, case number, exhibit number, and description.
- Collect a comparison standard.

Large Nonportable Objects Containing Paint Chips and Smears

- Scrape the paint fragments off one area, using a clean knife or scraping instrument.

- Remove the entire sample, getting down to bare metal or wood if necessary.
- Try to dislodge the fragments onto a clean piece of paper by tapping the object.
- Transfer the fragments from the paper into a pillbox, a glass vial, or other container that can be tightly capped. *Do not pack paint fragments in cotton. Do not allow paint to touch adhesives.*
- Seal containers and label with the date, initials, case number, exhibit number, and description.
- Collect a comparison standard.

Wet Paint Smears on Cloth, Wood, Metal, or Glass
- Let paint dry completely before placing the smeared item into a protective container.
- If possible, mark the item in an inconspicuous place with your initials.
- Place item in a container, and then seal and label the container with the date, initials, case number, exhibit number, and description.
- Collect a comparison standard.

Filings

- Collect filings caused by sawing, drilling, or filing by carefully lifting or scraping them into a plastic bag. *Use a nonmetallic device to transfer the filings to the collection bag.*
- Seal the bag and affix an evidence tag labeled with the date, initials, case number, exhibit number, and description.
- Collect a comparison standard.

Fragments

These materials come in the form of jimmies, fired bullets, fired bullet fragments, grillwork, headlight frames, dies, small tools, and so on. They can be loose or can be embedded in some matter.

Embedded Fragments

- Allow a laboratory to extract the fragments from the material in which they are embedded.
- Package as much of the solid matter holding the fragments as is practicable.
- Use a suitably sized container to hold the material.

- Seal containers after evidence has been placed in them.
- In all cases, either label the container or affix an evidence tag directly to the material with your initials, the date, and a case and exhibit number.
- Collect a comparison standard.

Loose Fragments
- Collect and place the fragments carefully in paper bags or other suitably sized containers. *Note: Each item should be packaged in a separate bag.*
- Package the items using padding that would prevent any damage to identifying characteristics or to the area along a fracture plane.
- Seal and label the evidence with the date, initials, case number, exhibit number, and description.

Large Sections
- Photograph visible marks on safe doors or other objects that are too bulky to remove.
- Preserve visible marks by casting them in a silicone rubber compound.

- Package and label the casting using the same methods employed for collecting impressions of footprints, but omit the "north" directional marking.
- Collect a comparison standard.

Paint
- Follow the procedures for viscous substances.

■ Small Objects

- At each crime scene, search for small objects, such as burned matches, particles of glass, broken fingernails, and cigarette butts.
- Follow the procedures outlined in this handbook for each of the known items. If you do not have specific directions for an item of evidence, place it in a crush-proof container without touching it directly with your fingers, and then seal and identify the container by writing on it in ink your initials, the date, and a case and exhibit number.
- Collect a comparison standard whenever possible.

Collection of Comparison Standards for Small Objects

Comparison samples of small objects or items found in the possession of a suspect or in his or her belongings should be submitted so that a comparison with items found on the crime scene can be made. Package and identify comparison samples as described earlier in the section "Small Objects."

■ Soil

Caked Mud

- Use a spoon, knife, or other instrument suitable for collecting pieces of caked mud. Any instrument used must be cleaned after each sample is taken.
- Place in a *clean* paper bag any personal articles, such as clothing and shoes, that bear traces of caked mud. Place each article in a separate bag.
- Seal each bag and label with the date, initials, case number, exhibit number, and description.
- Collect a comparison standard.

Dry Soil

- Collect at least half a pound of all available soil when possible.
- Place dry soil in a box or other similar cardboard container. *Do not use envelopes for dry soil. Do not use glass containers.*
- Seal and label with the date, initials, case number, exhibit number, and description.
- Collect a comparison standard.

Mud

- Use a *clean* knife and scrape mud off any objects that cannot be sent to the laboratory.
- Place scrapings in an envelope and then in an opened plastic or cardboard container for transportation.
- Seal the container and label it with the date, initials, case number, exhibit number, and description.
- Collect a comparison standard.

■ Wood and Fabrics

- Collect and package wood, carpeting, cloth, or other absorbent materials found near the origin of the fire that appear to contain traces of the accelerant or incendiary material. Use the same procedures described earlier for glass fragments.

■ Writing Instruments

- Handle the writing instrument with tweezers or cloth gloves, being careful not to smudge fingerprints.
- Dust for and collect fingerprints.
- Look for and submit instruments bearing teeth marks.
- Place instrument in a suitable crush-proof container and identify the container by writing on it in ink your the date, initials, case number, exhibit number, and description.
- Collect a comparison standard.

■ General Procedures

Use the procedures recommended and published by your agency for the transmittal of evidence and information to a laboratory for forensic examination. If you have no published procedures, use the following guidelines:

- After the evidence has been collected and preserved as described in Section II, package the evidence in a cardboard box whenever practical.
- Wrap liquid samples in padding to avoid breakage.
- Wrap the box with brown craft paper and seal the edges with either masking or paper tape, suitable for accepting a rubber stamp impression.
- Attach a transmittal letter to the package.
- Send the evidence package by registered mail or FedEx, UPS, or another delivery service that provides a signature-service capability or personally delivers it.

Word to the Wise

Paper envelopes and paper bags are the preferred way of submitting evidence in most cases. Paper allows the item to breathe, preventing mold and rust. This prevents contamination whenever the item is set down.

Notes

Notes

Notes

Notes

_____

_____

_____

_____

_____

_____

_____

_____

_____

_____

_____

_____

_____

Notes

Notes

_____

_____

_____

_____

_____

_____

_____

_____

_____

_____

_____

_____

_____

Notes

Notes

Also Available to Accompany the Evidence Collection Field Guide

Evidence Collection

Joseph J. Vince, Jr., William E. Sherlock, American Society for Law Enforcement Training

© 2005 • ISBN 0-7637-4787-4

$37.95 • Paperback

99 pages

Order your copy of *Evidence Collection,* by Joseph J. Vince, Jr. and William E. Sherlock today! The Evidence Collection handbook was developed with the special needs of both law enforcement officers and criminal justice students in mind.

In many areas of the country, responding officers may have to wait hours for laboratory resources to arrive at a crime scene—or they may never have the benefit of expert assistance at all. In addition, even major metropolitan departments have limited laboratory personnel that can be sent to crime scenes. Therefore, it is essential that field